edZOOcation™ presents:

Red

by Sara Karnoscak

Dedication:

For Miss Jackie. Thank you for helping all kids thrive.

−S.K.

For Mom & Dad , Tough like a Panda

−A.R.

Copyright © 2023 Wildlife Tree, LLC. All rights reserved.

Author: Sara Karnoscak

Designer: Allyson Randa

Editor: Tess Riley

Photo Credits:

AdobeStock.com

Pixabay.com

Pexels.com

ISBN: 979-8-9886164-2-9

This book meets **Common Core** and **Next Generation Science Standards.**

Table of Contents

4	A Red Panda's Body
6	Such Great Heights
8	Bamboo Thumbs
9	Piles of Poop
10	True Pandas
13	Other Names for Red Pandas
14	Too Cute!
16	Who Said It?
17	Animal Habitat
21	Animal Behavior
26	Food Web & Dangers
28	Can You Act Like a Red Panda?
30	Glossary
32	Jokes and Rhymes

A Red Panda's Body

Long wrist bone to grab bamboo.

Ankles that can turn backwards.

Bushy tail to balance and keep warm.

Two layers of rust colored fur.

Sharp claws for climbing.

Such Great Heights

Red pandas are some of the best climbers. They can even climb down trees headfirst! They turn their back feet all the way around. Then they can climb down upside down.

When they're in the trees, their bushy tails help them. They can move their tails to help them balance.

Smelly Pandas

Red pandas can smell with their tongues! They use the bottoms of their tongues to smell **odors** left by other pandas. Pandas leave odors when they walk. Their feet leave a special odor behind. The odor is strong to pandas, but humans can't smell it.

Odor: A Smell.

Bamboo Thumbs

Red pandas have a **false thumb** for grabbing bamboo. Their thumb is an extra finger made by their long wrist bone. Just like giant pandas!

Red pandas waddle. It's because they have front feet that turn inward.

False Thumb: An extra-long wrist bone on pandas that they use to grab bamboo.

Piles of Poop

Red pandas can poop their body weight in a week.

Male pandas will poop to let females know they are around.

They will also leave other odors for females. They wiggle their rear ends on things to rub a liquid for females to smell.

True Pandas

Red pandas are not closely related to giant pandas.

Giant pandas were named after red pandas. They were named after them because they both eat bamboo. And because they both have special "thumbs" for eating bamboo.

Scientists used to think red pandas were bears.

Then they thought they were raccoons.

Now they know they are in a family all their own.

Their closest relatives are skunks, raccoons, and weasels.

A Panda by Any Other Name

Can you guess which are real names for red pandas?

Lesser Panda	Firefox

Tree Hugger	Red Bear-Cat

Mini Panda	Red Racoon

Correct Answers: Lesser Panda, Firefox, Red Bear-Cat, Red Racoon

Too Cute!

Mothers make dens in hollow trees or **bamboo thickets**. They line them with moss and leaves.

Cubs are usually born in late spring or early summer. That's when there are the most tender bamboo shoots.

Cub: Baby panda.

Bamboo Thicket: A place where lots of bamboo grows.

Cubs are born with fur. They are about the size of a pickle. Their mother will clean them with her tongue, like a cat. She will carry them by their **scruffs**, also like a cat.

Scruff: *Back of the neck.*

Who Said It?

Red pandas are mostly quiet.

Sometimes they make a sound called a **huff-quack**. It sounds like a pig and a duck.

They can also bark, whistle, and twitter.

Huff-Quack: The sound a red panda makes that sounds like an oink and a quack.

Animal Habitat

Red pandas live in high mountains.

They live where there are trees and bamboo.

They spend most of their time in trees.

Red Pandas Live Where It's Cold. They Live In Asia

NORTH AMERICA

ATLANTIC OCEAN

PACIFIC OCEAN

SOUTH AMERICA

SOUTH ATLANTIC OCEAN

They live in the **Himalayas**.
They also live in other high mountains.

ARCTIC OCEAN

RUSSIA

EUROPE

AFRICA

ASIA

PACIFIC OCEAN

INDIAN OCEAN

AUSTRALIA

ANTARCTICA

Himalayas: *A mountain range with some of the world's highest mountains.*

Which Animal Can a Red Panda Sound Like?

PIG

CAT

DUCK

DOG

BEAR

MOUSE

Correct Answers: Pig, Duck, Dog

20

Sleepy Pandas

Red pandas sleep more than half the time. They are mostly **crepuscular**. That means they are most active at **dawn** and **dusk**. They are also active at night.

Crepuscular: Active at dawn and dusk.

Dawn: When the sun rises.

Dusk: When the sun sets.

Red Pandas Eat...

Bamboo

Fruit

Insects

Bird Eggs

Red pandas are classified as **carnivores**. But they hardly eat meat. They mostly eat bamboo. They can eat almost a third of their body weight in a day.

Carnivore: *Meat-eater.*

A Day in the Life...

As the sun starts to set, the red panda wakes.

He climbs down from the branch where he's been sleeping.

He grabs some bamboo and starts to munch.

# of a Red Panda

After he's eaten a belly-full, he climbs all the way to the ground. He needs to poop.

He poops where a female will find it. He also leaves his scent on a nearby tree.

Time to climb back to his sleeping branch. After a nap, it will be time to eat again.

The Red Panda Food Web

Red pandas eat a lot of bamboo. Sometimes they eat fruit, insects, and bird eggs.

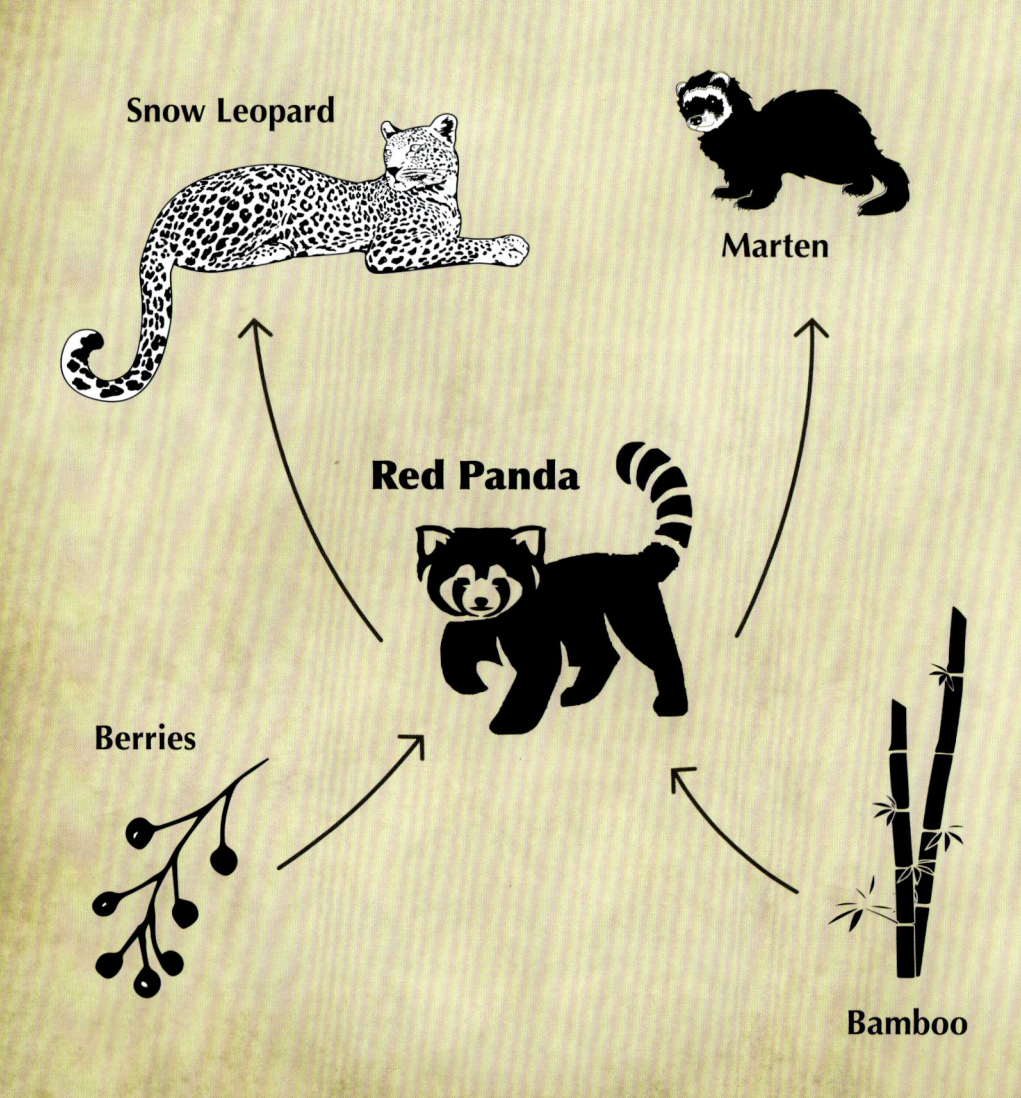

Snow Leopard

Marten

Red Panda

Berries

Bamboo

Dangers

Climate change is a danger to red pandas. Climate change makes them need to move to colder places.

Sometimes people take red pandas out of the wild. They sell them as pets.

Climate Change: The change in Earth's weather over a long time.

Can You Act Like a...

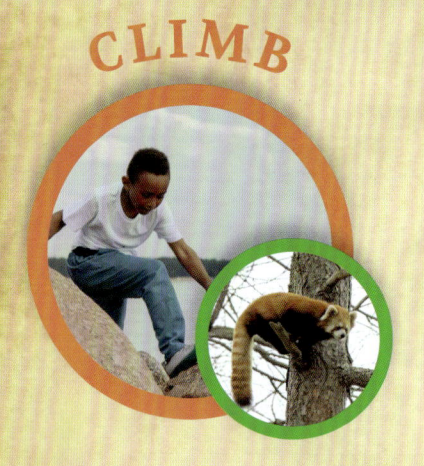

CLIMB

Red pandas are great climbers.

Find a safe place to climb.

EAT

Red pandas like to eat.

Eat your favorite healthy snack.

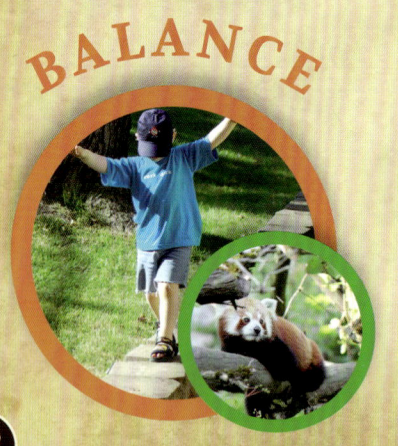

BALANCE

Red pandas have great balance.

Balance on one foot or on a small ledge.

Red Panda?

SOUNDS

Red pandas make funny sounds.

Can you make a huff-quack sound?

SLEEP

Red pandas like to sleep.

Curl up and rest in the coziest spot you can find.

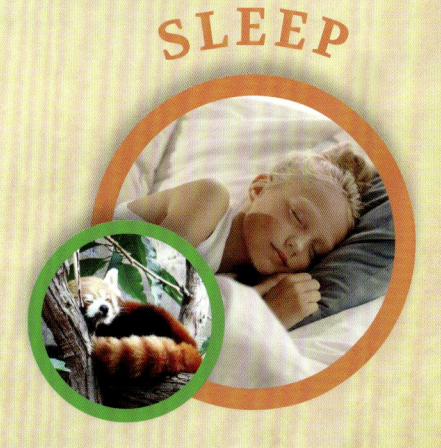

Glossary

Bamboo Thicket: A place where lots of bamboo grows.

Carnivore: Meat-eater.

Climate Change: The change in Earth's weather over a long time.

Crepuscular: Active at dawn and dusk.

Cub: Baby panda.

False Thumb: An extra-long wrist bone on pandas that they use to grab bamboo.

Dawn: When the sun rises.

Dusk: When the sun sets.

Huff-Quack: The sound a red panda makes that sounds like an oink and a quack.

Himalayas: A mountain range with some of the world's highest mountains.

Odor: A smell.

Scruff: Back of the neck.

Jokes and Rhymes

Every visitor at the zoo
Always stopped to look and say, "Ooo!"
The red panda looked so cute
As he ate his bamboo shoots
That even the birds stopped to coo.

Did you hear about the party at the zoo?
It was panda-monium!

What do you call a red panda that doesn't want to grow up?
Peter Panda.

What does a ghost panda eat?
BamBOO!

Why is panda one of the easiest words to spell?
To spell it, all you need is **P and A**.

P AND A

Why do pandas have fur coats?
Because they'd look silly in denim jackets.

What is a red panda's favorite breakfast?
Panda-cakes.